EDUCATION
KT-425-360

2100063520
E.L.R.S.

Rubbish

Written by Sally Morgan

Wayland

Bodies	Fairgrounds	Light	Special Days
Boxes	Growth	Patterns	Textures
Changes	Holes	Rubbish	Weather
Colours	Journeys	Senses	Wheels

Picture acknowledgements
The publishers would like to thank the following for allowing their photographs to be reproduced in this book. Bruce Coleman Ltd. 4 (below), 6 (above/Norman Tomalin), 13 (Carol Hughes), 16 (Jane Burton), 18 (above/Hans Reinhard), 18 (below/Leonard Lee Rue), 19 (Christian Zuber), 21 (below/William S. Paton), 22 (above/George McCarthy), 22 (below/Chris Catton), 24 (Chris James); ECOSCENE 5 (Morgan), 6 (below/Cooper), 7 (Morgan), 8 (Harwood), 9 (Adrian Morgan), 10 (above/Hawkes), 10 (below/Greenwood), 11 (above/ Morgan), 11 (below/Wilkinson), 12 (above/Morgan), 12 (below/Gryniewicz), 14 (Morgan), 15 (above/Cooper), 15 (below/Wilkinson), 20 (Morgan), 21 (above/Wilkinson), 23 (Morgan), 25 (above/Lees), 25 (below/Wilkinson), 26 (above/Tweedie), 26 (below/Gryniewicz), 27 (above/Gryniewicz), 27 (below/Tweedie), 29 (above/Morgan), 29 (below/Brown); Oxford Scientific Films Ltd. 17 (above/Jack Dermio), 17 (below/Alastair Shay); Wayland Picture Library *title page*, 4 (above), 28.

Cover photography by Daniel Pangbourne, organized by Zoë Hargreaves. With thanks to the Fox Primary School. A special thank you to Zinka and Ordrej.

First published in 1993 by
Wayland (Publishers) Ltd
61 Western Road, Hove
East Sussex BN3 1JD, England

Editor: Francesca Motisi
Designers: Jean and Robert Wheeler

Consultant: Alison Watkins is an experienced teacher with a special interest in language and reading. She has been a class teacher but at present is the special needs coordinator for a school in Hackney. Alison wrote the notes for parents and teachers and provided the topic web.

British Library Cataloguing in Publication Data
Morgan Sally.
Rubbish. – (Criss Cross Series)
I. Title II. Series
363.72

ISBN 0-7502-0767-1

Typeset by DJS Fotoset Ltd, Brighton, Sussex
Printed and bound in Italy by L.E.G.O. S.p.A., Vicenza

Contents

Words that appear in **bold** in the text are explained in the glossary on page 32.

What is rubbish?

We throw away all sorts of things – old newspapers and magazines, plastic bags, cardboard boxes, uneaten food and things we just don't want any more.

How many different types of things have been thrown away in this picture?
Do you throw away anything else?

Collecting rubbish

Our rubbish is taken away by a dustcart. We put out our bins and the dustbin men and women collect our rubbish and put it into their dustcart.

How much rubbish is collected from your home?

Buried or burnt?

Do you know what happens to your rubbish after it is collected? In many places, the rubbish is taken to huge holes in the ground and buried. These are called landfill sites. There are lots of birds around these sites looking for food amongst the rubbish.

Sometimes the rubbish is not buried, but it is taken to special places where it is burnt instead. The heat from the burning rubbish can be used to make **electricity**, as at this **power station** in the USA.

Nature's rubbish

Plants produce rubbish too. In autumn, many trees drop their leaves. They fall off and collect on the ground. This is called leaf litter and it is the rubbish of the trees.

Worms in the soil pull the leaves into their burrows, while ants and beetles carry away small pieces of leaves.

There is little left of the leaf litter by the following summer.

Nature's cleaners

Some animals are very important as they clean up after other animals. In Africa vultures feed on the dead bodies of animals.

Smaller animals are cleaners too. This dead **chameleon** is soon picked clean by an army of ants.

These **dung** beetles are feeding on the droppings produced by larger animals. They roll up small pieces of dung into a ball, so that it is easy to push home.

Then they bury it and lay an egg in it.
When the young dung beetle hatches it will feed on the dung.

Dropped on the ground

Too many people drop their litter on the ground. Litter soon builds up in cities. The children on this bench are eating and drinking.
What has happened to some of their litter?

When people drop their litter, other people have to clear it up.

Rubbish and animals

In many countries animals come into towns and cities to look for food. They think they will find a free meal, but they can bring diseases with them and cause lots of damage.

In the USA racoons come into gardens and parks looking for food.

In Britain badgers, such as the one in this picture, sometimes steal food from our dustbins.

Hyenas, baboons and storks look for food in Africa.

This African elephant may be hurt by the cans on this rubbish heap. ▶

These polar bears in Canada are also looking for food amongst the rubbish.

Rubbish in the country

Many thoughtless people drop their litter in the country. Lovely places are spoilt by bottles, cans and food-wrappers dropped by walkers.

Rubbish spoils the country for people and wildlife too. These swans have built their nest surrounded by old tyres and other bits of rubbish.

Dangerous rubbish

Rubbish can harm animals as well as people. This mouse has become trapped in a bottle.

The duck has plastic rings stuck around its neck.

Broken glass is very dangerous. It could hurt people and animals very badly if it is left lying on the ground.

Recycling rubbish

We do not have to throw rubbish away. A lot of it can be used again, to make other things. It can be **recycled**. Most towns have special recycling points, where glass, paper, metal and cloth can be collected. These children in Sheffield are crushing aluminium cans for recycling.

What is being recycled in these pictures?

New from old

Many cans are made from aluminium. Aluminium is a metal that can be recycled over and over again. The old cans are collected and squashed together. They are then taken to a factory.

At the factory they are melted and made into brand new cans.

What do you do with your old cans? You can tell if a can is made from aluminium or steel by using a **magnet**. The magnet will not stick to aluminium cans.

Using rubbish again

Some people use their rubbish to make new things. This boy has made a model caterpillar from old egg boxes.

All the things in this photograph have been made from recycled paper. ▼

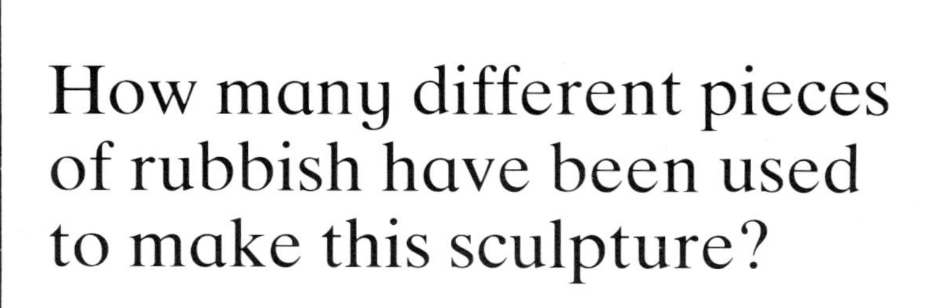

How many different pieces of rubbish have been used to make this sculpture?

Notes for parents and teachers

Science

• To encourage the children to think of their own and the school's use of materials they should think of ways of re-using everyday materials and objects, eg pots to store marbles and other bits and pieces in.
• The children should be aware that using materials wastefully is damaging to the environment and they should be encouraged to improve and protect the environment.
• To discover whether objects are biodegradable or not, bury a series of objects of the type normally thrown away. After 2-4 weeks investigate which materials show signs of breaking down and which could be composted.
• How dirty is your air? Design a fair test to discover where there was most air pollution.
• Rotting fruit is good for the ground and becomes soft and brown. Draw the changes you see everyday and keep a record.

Language

• An important way of raising the children's awareness of caring for our environment is through thoughtful discussion and relevant questioning by both child and adult. Language is a very powerful tool for changing and forming attitudes.
• Use of non-fiction and fiction books, eg *The Scrapyard Monster* by L. Anelli (Macdonald 1988). Also *Lucy's World* by S. Weatherill (Hippo Books 1991) brings environmental issues alive for 5-7 year olds through the eyes of Lucy Goose and friends.

Maths

• Rubbish audit. Make a list of all the things the class have thrown away in one week. Try the same exercise again making an effort to recycle or reuse as much waste as possible.
• Sorting leftovers of foods into, (a) things that rot and are good for the ground, (b) good for the birds, (c) thrown away.

Music

• Children can make up slogans, rhymes and songs on this theme. They can clap the rhythm and add simple percussion.

Technology

• Children can design their own packaging. Discuss what wrapping is actually needed, why we have layers of wrapping and their likes and dislikes.
• Design and build an animal's home out of rubbish (sticks, leaves, string, boxes, etc).

History and Geography

• Find out where the local recycling centre is and what can be recycled there.
• Looking for clues is an important historical skill. Children can investigate a rubbish bin and try and discover as much information about the person who threw the rubbish away.
• Children can go on a litter trail around the school or local environment. What types of litter do they find and where? Where are the dustbins placed?

Art

• The children can make a model from all sorts of rubbish. Can they recognize what has been used to make the model?
• Paint a big litter bin and stick things we don't want to recycle on the bin. How can we encourage people to stop dropping litter?
• Write to local scrap projects or industries to get unwanted or 'cut off' materials.

Dance/Drama

• Through movement children can explore the concept of 'cycles' and recycling as opposed to the waste chain.

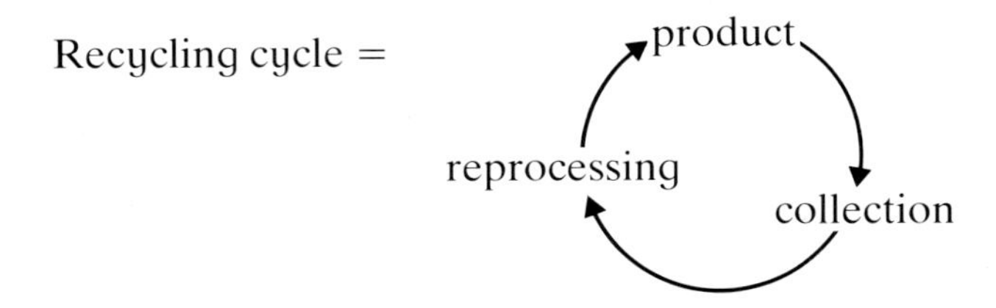

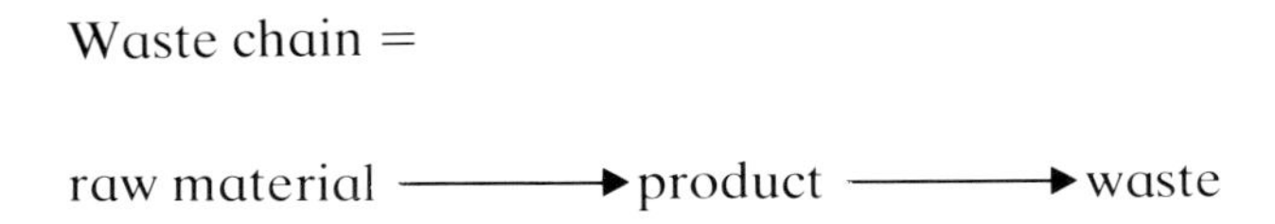

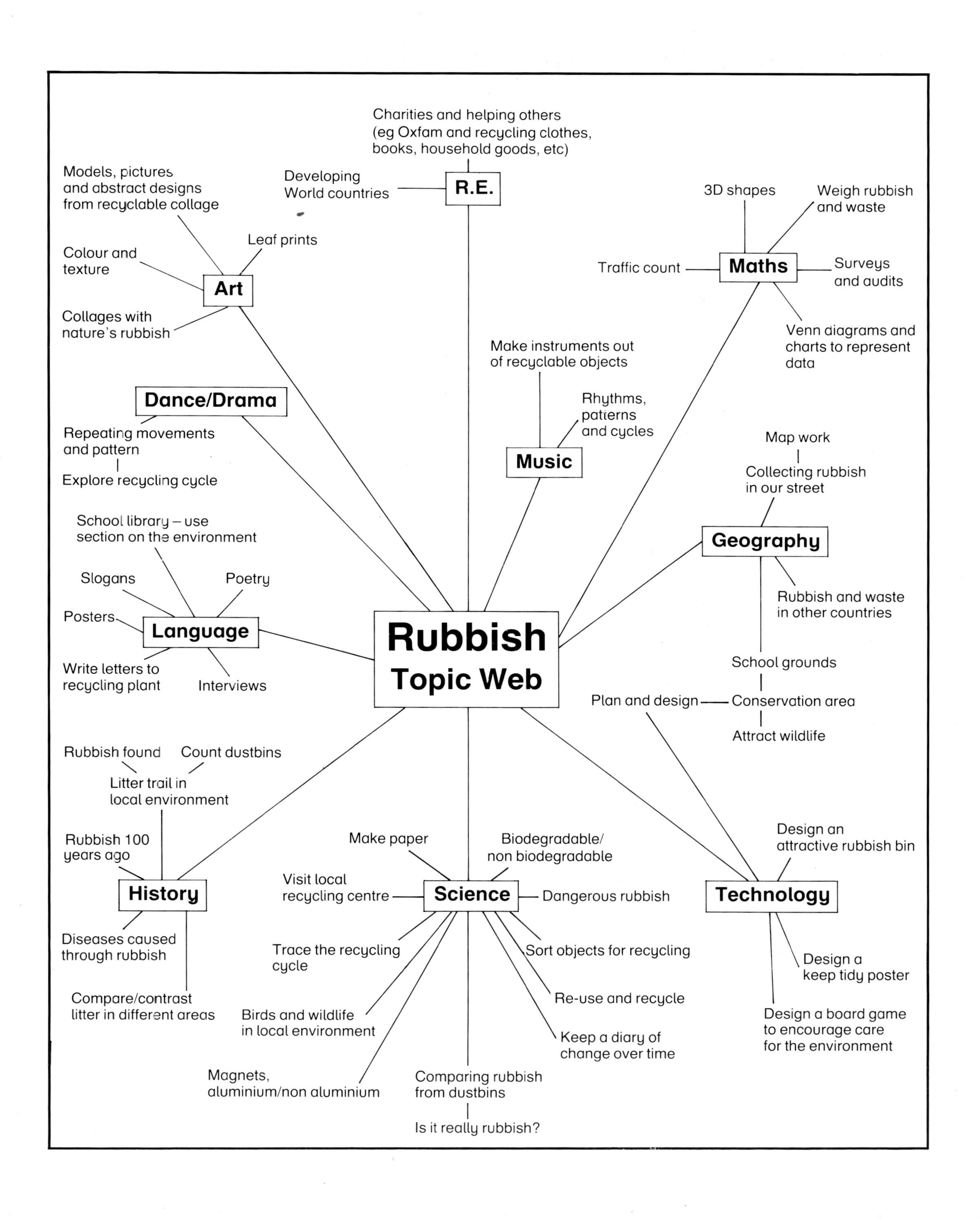

Rubbish
Topic Web
R.E.
Charities and helping others
(eg Oxfam and recycling clothes,
books, household goods, etc)
Developing
World countries
Art
Models, pictures
and abstract designs
from recyclable collage
Leaf prints
Colour and
texture
Collages with
nature's rubbish
Maths
3D shapes
Weigh rubbish
and waste
Traffic count
Surveys
and audits
Venn diagrams and
charts to represent
data
Music
Make instruments out
of recyclable objects
Rhythms,
patterns
and cycles
Dance/Drama
Repeating movements
and pattern
Explore recycling cycle
Geography
Map work
Collecting rubbish
in our street
Rubbish and waste
in other countries
School grounds
Conservation area
Plan and design
Attract wildlife
Language
School library – use
section on the environment
Slogans
Poetry
Posters
Write letters to
recycling plant
Interviews
History
Rubbish found
Count dustbins
Litter trail in
local environment
Rubbish 100
years ago
Diseases caused
through rubbish
Compare/contrast
litter in different areas
Science
Make paper
Biodegradable/
non biodegradable
Visit local
recycling centre
Dangerous rubbish
Trace the recycling
cycle
Sort objects for recycling
Birds and wildlife
in local environment
Re-use and recycle
Keep a diary of
change over time
Magnets,
aluminium/non aluminium
Comparing rubbish
from dustbins
Is it really rubbish?
Technology
Design an
attractive rubbish bin
Design a
keep tidy poster
Design a board game
to encourage care
for the environment

Glossary

Chameleon A lizard that can change colour to match its surroundings.

Dung Animal droppings.

Electricity A type of energy used for heating, lighting and working machinery.

Magnet A piece of metal that can pull iron or steel towards it.

Power station A place where electricity is made.

Recycled Used again and made into something else.

Index